Roshan   Cipriani

The New Dimensional Reality

Experiences Of Alternate Dimensions Of Existence

Reality Shifting

Table Of Contents

Dedication

To the song "***Bohemian Rapsody***" By _Queen_.

Freddie Mercury also known as _Farrokh Bulsara_,(a person like myself of Parsee descent),wrote it over several years using various styles of music. The song contains a segment which was very possibly inspired by Turiddo's aria "Mamma, quel vino è generoso," from _Pietro Mascagni's_ opera _Cavalleria rusticana_.

They together begin with the same word, they both are uttered in desperation by someone who has committed a crime and anticipates death, and they are both endeavoring to obtain a promise from a mother. They also both have the expression "...if I'm not back again," ("..._s'io non tornassi_".) The singer also bids the world goodbye proclaiming he has to go and gets ready to "face the truth" confessing "I don't want to die and I sometimes wish I'd never been born at all."

At its core the song "Bohemian Rhapsody" is about a young man who has inadvertently killed someone and, like _Faust_, sold his soul to the devil. On the evening before his execution, he cries to God praying, "_Bismillah_" (In the name of God in Arabic), and with angelic assistance, reclaims his soul from Shaitan/Satan or the devil in Arabic.

Recently (June, 2016) strange clouds were photographed over the CERN Large Hadron Collider (LHC) near the French/Swiss border at Geneva, Switzerland that many people believe shows that CERN is accessing another dimension by opening a portal in the sky or perhaps a parallel universe. That was the same day that CERN began a new experiment called "Awake." There are massive disturbances in the magnetosphere of the earth when the (LHC) Hadron Collider is turned on and it may have grave consequences for physical life on earth. It may also be a major impactor on the changing shifts in reality/dimensions for many people. There have been many voices in the scientific community that raised objections that CERN's LHC activities were unsafe, yet they were ignored. Have the scientist at CERN's LHC sold their souls? Is it time they faced the truth?

Only time will tell.

INTRODUCTION

Today everything began again. I looked around me and everything seemed grainy and disintegrating before my eyes.

Then I would blink, rub my eyes and it would go back to a kind of normal. Then there are those times when I wondered, "Am I crazy? Am I imagining the conversations I'm hearing? I am imagining the faces I'm seeing?"

I started sensing that something was very wrong and I needed to fix it. I started recognizing this, but I didn't know what any of it meant. I had no idea where to begin. Was I making this up somehow? Worst of all I didn't know that there were people having experiences like my own.

My journey has been challenging and often lonely, but for some unexplained reason I just seem to keep going. Recently it seemed to me that time was speeding up the days just seemed to fly by, and then they slowed to a crawl and it seemed as though one day took the time of two days to finally pass. Just what is actually happening?

This book has been written to offer a different view on reality than the norm.

Quantum Mechanics views parallel universes as a series of possibilities, so we accept that. However today we understand that we live in three separate and distinct realities, our physical world, our mental world and our dreaming sleep state world. In the dreaming sleep state world we exist there for almost one-third of our lives, so it is quite significant. Is it also possible to travel to a parallel universe to change things around that you do not want or like in your physical world? I believe it is quite possible.

This is why I believe sometimes obvious end results of certain activities did not result in their predicted endings. There was a perceivable moment in which everything changed. Things looked much different before that time, even the sunshine was a lot more golden yellow. In point of observance most colors looked completely different; they contained more warmth somehow.

In interactions with individuals I found that many people were also completely altered. How else can you explain different outcomes, different subtle little changes that just don't quite add up? It often feels as though this is an alternate reality and there is nothing I can do to return to the one I am familiar with. Life has become so strange and the implications are so upsetting to me that I try not to think about it unless I am once again faced with another example that I just cannot ignore. Usually it is not until I get home and sit and reflect on the day that a strange eerie feeling hits me. This feeling is best described as if something important has somehow changed, usually both my daughter and I have experienced it and it makes us sometimes question our reality.

Since Quantum Mechanics states that every eventuality is played out somewhere in the universe, we could have entered into an area with different eventualities. Essentially I sometimes would rather not think about the reality of this if it's true, since that would open up many more questions that would demand answers and right now I don't have any. I often think to myself that if I went somewhere else the people that I loved who are dear to me must have come along as well and perhaps they are not able yet to notice the little changes. The alternate thought is just to upsetting so I rarely consider it. What if I alone am somehow into another dimension? No, you can't simply move to another reality and everything continues as before.

I would love to find a different logical explanation, and so I take comfort in the idea of decisions.

Perhaps the different choices we have made throughout our lives, are the outcomes here and yet by choices with different outcomes in other dimensions we notice differences somehow only when we are transported there.

There is consciousness in things and areas in which perhaps we have never considered. Could this be impacting our reality. The basic structures of our own brains are simply bands of energy. This energy is the basis of everything. Modern Quantum physics asserts that this energy is not dormant or silent it is pulsating and it is aware. The ancients knew of this energy and had many names in various cultures. Even this computer that I am writing this book on is filled with a form of conscious life. Seems farfetched to you? Think again.

The Global Consciousness Project, also termed the *EGG Project*, is a global, multi-faceted partnership of researchers, engineers, and artists. They have collected information uninterruptedly from an international set-up of physical random number generators situated in 65 host locations across the earth. The outcome is a record of coordinated parallel arrangements of random numbers. The information is archived on a server at Princeton University and subjected to strict examination and testing to learn whether there are departures from predictable randomness consistent to alterations in worldwide events.

Eventually Princeton University decided to carry this experiment a bit further. They placed an observer and asked them to hold the thought of the machine changing its random generation, and the machines actually did.

They have actually discovered that the greater the event in our global awareness for at least a half hour or more; as a shared collective experience, the greater the change in the data. During moments of significance to humans, the devices displayed slight parallels with each other. The effects are greater in ratio to the prominence of the actions they studied, and they were higher if the level of emotional participation is elevated. The conclusion is that something connected with mass consciousness is capable of altering the physical world. This is evident from the alterations on the number generators. Yes, scientific research now suggests human consciousness can influence physical reality. The potential for consciousness to affect chaotic systems is termed mind-matter interaction.

Are we all connected in some way that we don't fully understand but that is measurable in some way? God is the ultimate planner and today we are at a precarious time in history. We face the inevitability of transforming our development as people into one that is more conscious of capabilities and responsibilities. Our designer already has all of this already built in. Recent studies have shown that when we actually change our mind, there is a physical proof of change in the brain itself.

University of London physicist David Bohm, for instance, believes that objective reality may not be existent, that notwithstanding its seeming solid nature the universe is at its core an illusion, an enormous and superbly complete hologram. A hologram is a three-dimensional photograph made with the assistance of a laser. If a hologram of a flower is cut in half and then lit up by a laser, each half will still be found to contain the entire image of the flower.

Undeniably, even if the halves are divided again, each piece of film will always be found to have a reduced but complete version of the original image. Unlike standard photographs, every fragment of a hologram comprises all the information possessed by the entire item and appears to be the item in complete form.

Bohm also considers the reason subatomic particles are able to remain in communication with one another irrespective of the expanse separating them is because their separateness may be an illusion.

He contends that at some profounder level of reality such particles are not individual entities, but *are really extensions of the same central something*. According to Bohm, the seeming faster-than-light communication between subatomic particles demonstrates a deeper level of reality and a more complex dimension beyond our own that is unknown to us.

So yes, if there are multiples versions of you, you could seem to be the complete version, while only being a part of a greater mass of the subatomic particles that make up yourself. If the apparent separateness of subatomic particles is deceptive, it means that at a deeper level of reality all things in the universe may be infinitely interconnected.

Stanford neurophysiologist Karl Pribram has also become convinced of the holographic nature of reality. Pribram believes memories are programmed not in neurons, or minor assemblages of neurons, but in designs of nerve impulses. These impulses network the whole brain just like patterns of laser light interference intersect the complete piece of film holding a holographic picture.

Pribram believes the brain is a hologram and his view has increasing support among fellow neurophysiologist. Even dreams and experiences involving "non-ordinary" reality can be understandable under the holographic example.

What this means for my theory is that in a universe in which individual brains are actually portions of a greater "hologram" and where everything is infinitely interconnected there could be multiple versions of ourselves with access to memories of the other versions of ourselves. Scientist state that there are parallel dimensions existing and this could be where there are multiples of each person scattered throughout.

I think we now are constantly being sucked in and out of dimensions and we will do not get back in to our original dimension. This would explain having clear memories of things that never occurred in the dimension/reality you are standing in at present.

Something unexplainable has happened. For numerous people the world seems to have changed. Perhaps when a dimension/ reality becomes discordant with your continued existence; you die there, then your consciousness "merges" to the adjacent useable dimension/reality in which you are still alive, and you continue to live on as you but with subtle differences. A new side effect of this is that the new dimension/reality is actually inconsistent with your memories, because the reality/ dimension that was completely constant with your memories is the one you left because you died there.

In that reality/dimension you would have died and could no longer remain there and so scientifically speaking you cannot stay to observe your own demise. According to my understanding of particle physics, no self-observing system can observe its own collapse without the observer affecting or altering the outcome.

There is a theory called Quantum Immortality. The notion being, if we die in one reality/dimension/time stream, we may just fuse into an alternate version of ourselves in another reality/dimension/time stream.

The way I see it, we seem to be able to shift in between different dimensions/realities and they don't seem like they are correct or "normal" somehow to our own experiences. Although we have a very powerful and recognizably magnificent personal instrument; our Brain, which we still do not completely understand, we are willing to assume the "mistake" must be with us, and our thinking on some massive scale.

The world is a big, mysterious place, and it'd be foolish to believe we have all the answers, and now only these two choices. Our understanding is limited, but many of us who have experienced these things know something is radically altered and different.

Recently scientist proposed the theory called the "many interacting worlds" hypothesis (MIW). This theory confirms the idea that parallel worlds don't just exist, but they also intermingle with our very own DAILY. However they believe that parallel worlds only interact with our world on a quantum level and therefore are not easily detectable.

The Mandela Effect is a theory of parallel universes, constructed within the idea that many people have similar alternative memories about past events, and they may all have been in a different dimension with different occurrences and outcomes and not be mistaken in their recollections. The Mandela Effect was first designated as such online in 2008/2010. A woman named Fiona Broome discovered that others had a distinct and different memory from the norm, and it was similar to hers. She along with many other people remembered that Nelson Mandela had died during his imprisonment in the 1980s and was confused as to how he could have now died again in 2013. This revelation opened the conversation and people began revealing other strange recollections not always shared by people around them, yet something they have in common with many other people in the world.

If you haven't experienced the Mandela effect or don't react to it as I have, this may not make much sense to you. Consciousness, the state of being self-aware or awake is something physicists have not come into agreement on concerning a descriptive theory but have continued to delve into.

Quantum Brain Dynamics (QBD) is the theory that states there is a new quantum field that is accountable for consciousness. If this is true, it would signify that our personal consciousness survives by means of communication with a quantum field, or due to contact by quantum particles. We know that all possible arrangements of particles exist at all times, and we see results from that arrangement, however something is strange. The news stories of deaths that do not apply anymore and other strange occurrences make one stop and question if our conceptual abilities are being tampered with.

Having experienced other alternate memories and realizing that I wasn't alone – and that many other people have shared the same alternate memories was just the starting point in my research. However I believe that someone has altered something that has impacted our universe and apparently other dimensions as well and they have certainly received our attention.

Chapter One

The Word In Your Own Voice

If this phenomenon lasts or increases a lot more people are going to find themselves in a world they no longer are familiar with. For instance, what if you wake up one day and the country you lived in no longer exists but you have clear memories of being there your whole life. What if your spouse or parents don't even remember there was such a place? Do you just accept their memories as more valid than your own? Do you just forget what you KNOW since you cannot explain it even to yourself? The idea is actually quite frightening. The idea is actually becoming a daily reality for many people.

I've tried to think of rational explanations and it is very difficult. Assuredly remembrances can be flawed but when they are personal and you can recollect the specifics of exactly where you were, what you were doing, and even thinking at the time these memories were made, it's difficult to just discount them as nothing important, or that you are mistaken.

There is however one thing that I have noticed that seems fairly consistent in all this. It is that people like me seem to constantly be told that we look "young for your age." I am 56 years old and people always think on meeting us that my daughter (33) and I are two friends or sisters, which is usually very flattering to me. She is also told the same thing and people are surprised that she is over 25. I know several people that I have recently met that look 10-15 years younger than their actual age. Could our original dimension/reality have stopped time at some point temporarily and it is reflected in our appearance.

Since we know that people's ages are determined by body and biochemical progressions in our physiques, undoubtedly people would age more slowly at relativistic speeds. Particles surely do, because particle accelerators produce some short lived particles, such as muons or pions, and they can survive for much longer than their lifespan would be when at rest in the laboratory situation. It seemed to me recently that time just slowed to a crawl and a few days later it just sped up so that a whole day passed by unbelievably faster than I thought possible. The people I have known for years would probably think I was crazy if I started speaking to them about some of the things I am writing about now. It has taken me a long time to even be able to write down the things I started noticing since 1999, for fear of being laughed at. At one point I actually thought I might be losing my mind, that however has changed and I now know this is quite real.

It recently took a more serious turn for me as I am a primarily biblical writer. Having grown up in a fundamentalist Christian house hold, if there is one thing I am very familiar with; it is my Bible. I have found that there are very big changes to what I know as my memorized scriptures. Just by looking at the scripture on Bible sites on-line and by checking in my own personal copy of my Bible (owned over 20+ years) the changes are unbelievably reflected there as well. When I reference changes I mean the changes are to my physical KJV Bible. In this dimension/Reality these things are completely different.

I remember clearly that in Mathew 6:10 the words were:

" Thy will be done **on** Earth, as it is **in** heaven."

In Matthew 6:12 the words were: "And forgive us our **trespasses,** as we forgive those who **trespass against us**."

Our family only used the King James Version of the Bible so that is what I memorized and it said "Trespass" not "Debt" which it now says in this reality/dimension version of the KJV.

Another change is "And forgive us our **debts**, as we forgive **our debtors**." Matthew 6:12. The KJV now says Debt. I know it did **NOT** in my prior experience reality/dimension.

The Scriptures in every King James Bible whether on-line or in print seem to now have been altered worldwide or in just about each language of this completely different reality/ dimension be completely changed from what I have known.

Other changes I have noticed and learned that others have seen as well are these:

Luke 17:34: "two people in one bed," or "two in one bed", or, "**two men in one bed**"?

Matthew 24:41: "**t**wo women grinding at the mill" or "**two grinding grain at the mill**"?

Matthew 2:17: "**Prophet Jeremiah**," or "Prophet Jeremy"?

Luke 19:23: Wherefore then gavest not thou my money "**into the bank,**" that at my coming I might have required mine own with usury or "**to the exchangers**"?

There definitely was no mention of "a bank" in the KJV.

Mark 2:23: And it came to pass, that he went through "**the corn fields**" on the sabbath day; and his disciples began, as they went, to "**pluck the ears of corn**" or "**grain fields**" and "**plucked the grains**"?

Luke 6:1 And it came to pass on the second sabbath after the first, that he went through "**the corn fields**"; and his disciples "**plucked the ears of corn**" and did eat, "**rubbing them in their hands**" or "**the grain fields,**"? Why would you rub corn in your hands?

Luke 17:31 In that day, he which shall be upon the housetop, and "**his STUFF**" in the house, let him not come down to take it away: and he that is in the field, let him likewise not return back.

Luke 17:31 "In that day, he which shall be upon the housetop, and *"his possessions"* in the house, let him not come down to take it away: and he that is in the field, let him likewise not return back."

I know that the word *"stuff"* was definitely **NOT** there previously.

The Bible said *"the lion would dwell with the lamb"*. The phrase *"the lion will lie with the lamb"* has a completely different connotation. It means that they would mate, which is forbidden in scripture.

Isaiah 11:6 has been altered to "wolf" and the theme of it has been completely removed now "the wolf is going to dwell with the lamb."

 I know that it did not state wolves as the connotation for wolves was always negative.

Matthew 7:15 K Beware of false prophets, which come to you in sheep's clothing, but inwardly they are ravening *wolves*.

I recently read the account of someone so distraught over this that they visited several different churches and spoke to numerous people; some actually working at the church offices. While many of the pastors themselves did not seem alarmed (no surprise there) all of them including secretaries and other church support staff had no memory of "the wolf will dwell with the lamb" but they did remember "the lion will dwell with the lamb."

Luke 19:27 now says "But those mine enemies, which would not that I should reign over them, bring hither, *"and slay them before me."* *Jesus taught* *"love your enemies"* *but now it says he said* *"slay them before me."*

Matthew 9:17 now says "Neither do men put new wine into *"old bottles: else the bottles break"* and the wine runneth out, and "the bottles perish": but they put new wine "into new bottles," and both are preserved. The KJV *never* used the word Bottles. It was wineskins.

I memorized Psalm 46:10 it said: "Be still, and know that I am God: I will be exalted *among the nations*, I will be exalted in the earth."

This has been changed to "Be still, and know that I am God: I will be exalted *among the heathen,* I will be exalted in the earth"

I have read the entire Bible, including both Apocrypha many times in thirty years. I have also studied various translation of the Bible due to my work and research. I have read all of the Old Testament and pseudepigrapha, apocrypha and sacred writings along with most of the surviving New Testament apocrypha and the gnostic gospels including the Lost Books of the Bible. ***Nothing could have prepared me for this drastic change in what I knew as the bible.***

There are now multiple verses in the Old Testament using the word "matrix," in place of the original word "womb." In this dimension/reality they did not however replace all instances of the word womb in scripture.

*Exodus 13:12 That thou shalt set apart unto the Lord all that openeth the **matrix,** and every firstling that cometh of a beast which thou hast; the males shall be the Lord's.*

*Exodus 13:15 And it came to pass, when Pharaoh would hardly let us go, that the Lord slew all the firstborn in the land of Egypt, both the firstborn of man, and the firstborn of beast: therefore I sacrifice to the Lord all that openeth the **matrix**, being males; but all the firstborn of my children I redeem.*

*Exodus 34:19 All that openeth the **matrix** is mine; and every firstling among thy cattle, whether ox or sheep, that is male.*

*Numbers 3:12 And I, behold, I have taken the Levites from among the children of Israel instead of all the firstborn that openeth the **matrix** among the children of Israel: therefore the Levites shall be mine;*

*Numbers 18:15 Every thing that openeth **the matrix** in all flesh, which they bring unto the Lord, whether it be of men or beasts, shall be thine: nevertheless the firstborn of man shalt thou surely redeem, and the firstling of unclean beasts shalt thou redeem.*

*Genesis 3:15 **"crush"** and **"bruise"** "And I will put enmity between thee and the woman, and between thy seed and her seed; (he instead of it?) it shall **CRUSH** thy head, and thou shalt **BRUISE** his heel."*

This is now changed to **"bruise" and "bruise"** *"And I will put enmity between thee and the woman, and between thy seed and her seed; it shall* **BRUISE** *thy head, and thou shalt* **BRUISE** *his heel."*

Luke 20:24**"Denarius" to "Penny"** "Shew me a denarius. Whose image and superscription hath it? They answered and said, Caesar's."

This is now changed from **"denarius"** to **"penny"** "Shew me a penny. Whose image and superscription hath it? They answered and said, Caesar's."

This verse in Revelation is also changed to **"penny"** from **"denarius."**

Revelation6:6"And I heard a voice in the midst of the four beasts say, A measure of wheat for **a penny**, and three measures of barley for **a penny**; and [see] thou hurt not the oil and the wine."

Mark 13:10**"preached"** to **"published"** "And the gospel must first be **"PREACHED"** among all nations."

This verse is now also changed. "And the gospel must first be **"PUBLISHED"** among all nations."

Matthew 26:45 Jesus commands them to wake up." Then cometh he to his disciples, and saith unto them, **"awake, from your rest"** behold, the hour is at hand, and the Son of man is betrayed into the hands of sinners."

This verse has been changed to Jesus commanding them to sleep.

"Then cometh he to his disciples, and saith unto them, **"Sleep on now, and take your rest"**: behold, the hour is at hand, and the Son of man is betrayed into the hands of sinners."

Luke 22:20 Changed from **"New Covenant"** to **"New Testament"**

Luke 22:20 Likewise also the cup after supper, saying, This cup is the **new covenant** in my blood, which is shed for you

Now it states in Luke 22:20 Likewise also the cup after supper, saying, This cup is the **new testament** in my blood, which is shed for you.

Acts12:4 Changed from **"the Passover"** to **"Easter"** in the KJV.

Now it states Acts 12:4 "And when he had apprehended him, he put him in prison, and delivered him to four quaternions of soldiers to keep him; intending after **Easter** to bring him forth to the people."

BEFORE Acts 12:4 "And when he had apprehended him, he put him in prison, and delivered him to four quaternions of soldiers to keep him; intending after **the Passover** to bring him forth to the people."

BEFORE John 3:16 KJV "For God so loved the world, that he gave his only begotten Son, that whosoever believeth in him **SHALL** not perish, but have everlasting life."

AFTER John 3:16 KJV "For God so loved the world, that he gave his only begotten Son, that whosoever believeth in him **SHOULD** not perish, but have everlasting life."

BEFORE Luke 21:9 "But when ye shall hear **of wars and rumours of wars**, be not terrified: for these things must first come to pass; but the end is not **forewith**"

AFTER Luke 21:9 "But when ye shall hear **of wars and commotions**, be not terrified: for these things must first come to pass; but the end is not **by and by.**

BEFORE Genesis 9:16 "And the **RAINBOW** shall be in the cloud; and I will look upon it, that I may remember the everlasting covenant between God and every living creature of all flesh that [is] upon the earth."

AFTER Genesis 9:16 "And the **BOW** shall be in the cloud; and I will look upon it, that I may remember the everlasting covenant between God and every living creature of all flesh that [is] upon the earth."

BEFORE Galatians 4:25 "For this **HAGAR** is mount Sinai in Arabia, and answereth to Jerusalem which now is, and is in bondage with her children."

AFTER Galatians 4:25 "For this **AGAR** is mount Sinai in Arabia, and answereth to Jerusalem which now is, and is in bondage with her children.

My memory is that the Book of Revelation ended with Revelation 22:20, which is: "He which testifieth these things saith, Surely I come quickly. Amen. Even so, come, Lord Jesus."
Now there is a **NEW VERSE ADDED** Revelation 22:21 "The grace of our Lord Jesus Christ be with you all. Amen." (Protestant Benediction ?)

BEFORE Ephesians 3:19 "Now unto Him who is able to do exceedingly abundantly above all that we ask or think, according to His power that worketh in us. 20 **To Him be the glory, honor, and praise forever and ever. Amen"**

AFTER Ephesians 3:19 19 "Now unto him that is able to do exceeding abundantly above all that we ask or think, according to the power that worketh in us, **20 Unto him be glory in the church by Christ Jesus throughout all ages, world without end.** Amen"

The Catholic word ***BISHOP*** has now replaced ***ELDER*** in the KJV.

1 Timothy 3:2 "A ***bishop*** then must be blameless, the husband of one wife, vigilant, sober, of good behaviour, given to hospitality, apt to teach;"

Acts 1:20 "For it is written in the book of Psalms, Let his habitation be desolate, and let no man dwell therein: and his ***bishoprick*** let another take." ***Bishoprick replaced office***.

Philippians 1:1 "Paul and Timotheus, the servants of Jesus Christ, to all the saints in Christ Jesus which are at Philippi, with the ***bishops*** and deacons:"

1 Timothy 3:1 "This is a true saying, if a man desire the office of a ***bishop***, he desireth a good work."

Titus 1:7 For a ***bishop*** must be blameless, as the steward of God; not selfwilled, not soon angry, not given to wine, no striker, not given to filthy lucre;

1 Peter 2:25 For ye were as sheep going astray; but are now returned unto the Shepherd and ***Bishop*** of your souls.

The simple fact IS that the literal words have changed _in our own books in this reality._ I believe that the changes to the King James Version of the Bible started to appear sometime in the last eight months so a major shift must have occurred at that time. I also believe that the changes that have produced this were not recent but made a long time ago. We have moved into this time in this reality/ dimension now without most people really noticing anything different or being aware of the relocation unless they had memorized the Biblical words and to their shock found them changed from what they have known.

What I clearly remember regarding these examples does not exist in this present reality/ dimension. A single word or often the position of words has changed. If you searched the scripture sites on-line or searched in any Bible-- even your personal copies of many years, the scriptures are completely different.

I have an excellent memory and I know for certain that many things are clearly different here.

I personally believe there is something tremendously powerful and treacherous at work. Whoever chose these items to alter has betrayed themselves by the very choices that they have made. Only someone with a surface or passing acquaintance of Biblical scripture would have needed to choose those oft mentioned texts.

They either had no idea how many people would have eventually studied the original and *memorized the King James Version* of them and would instantly KNOW that the books wordings are completely different from what they knew, **or that is exactly what they were counting on to get our attention.**

It seems as though only printed copies, facsimiles, and mechanically produced copies such as from copier machines are different. So far words that are handwritten seem to remain as they were executed in our old remembrances.

It's also noticeable that only the King James Version of the Bible itself had been changed, while the many documents written citing the identical words remain untouched. It's almost as if the persons orchestrating this just set out to pick passages that they knew people would take exception to having the wording changed. So this is either an effort to discredit an ancient book and make the words seem silly and irrelevant, *they have actually done this to get our attention.* They have also let us know by doing this that yes, we are *either somewhere else* and something very real is going on or our old reality has been altered.

Dishonesty is not perfect. It can never be, and it once again has betrayed itself.

Reckless people with strange equipment and technology used for wickedness with an assignment to put an end to present reality would go after the Biblical texts, but what if they have a bigger agenda? Perhaps what they have caused to happen is even greater than that.

However, someone who wanted to signal the world that the reality they may be in is not their original home, could have chosen no better flag to hoist than altering the King James Version of the bible. This may have been done when our old reality merged with other time /reality/ dimensions in which the altered biblical text had already been in existence there. This dimension would have to be much further away from our original one, for there to be so many obvious differences that are easily noticed.

For people who don't know that this is a different place, just reading and seeing the bible scripture changes here is a major shock. Those that do see what has happened understand that the things in this reality weren't just changed or altered , *but that we are probably from a parallel reality/dimension that has some major differences from what we have known in the Bible.*

Well, this is where having the scriptures stored in your memory becomes invaluable. Yes, it is the word in your own voice that you will have to rely on for your understanding now in this place.

The idea that there could be multiple versions of us is easier to accept than multiple version of the KJV Bible for many people. The idea that there has been some strange paradigm shift and the world you think you were living in and the information that you embrace as fact was not the only fact but another version of those facts.

Additionally it takes a large leap forward in thinking to consider that the fabric of reality shifted at some point in the past, and we are in a parallel inhabitable reality/dimension/ universe. It even seems that we are continually moving between them. Yet the question remains, could this all be due to the experiments of time bending that may be happening repeatedly in the Large Hadron Collider located at CERN? Maybe there are other technologies working in tandem with them as well.

The way I see it the only proof we have, if WITHOUT A DOUBT WE ARE IN A NEW REALITY/DIMENSION/UNIVERSE, are the memories of those of us who for whatever reason shifted to this reality/earth/dimension/universe, and kept those old memories.

A theory is a hypothesis pending evidence. When evidence challenges a theory, it's time to change the theory, not time to change the evidence to fit the hypothesis. For me the massive amount of remembered information that is significantly different from this current present reality dimension is sufficient evidence to support the hypothesis that we are no longer in our original dimension/ reality. Although this statement sounds wild, it could have some enormous implications for every person alive here today.

I believe that sometimes when we dream, our dreaming actually is the experiences of our alternate selves in another dimension/reality. Perhaps this may explain waking up from an eight hour nights rest totally exhausted. Perhaps dreaming is seeing the other lives through your other eyes in your dreams.

Then there are the very obvious changes to vital life facts that you find here in this dimension /reality. The Universal Blood Donor is now O Rh Negative, where before as I knew of this it was O positive. I remember this because I always thought I was special and could help anyone. Now I wonder what my blood type in this reality/dimension would actually be. In the past for me very few people were RH Negative and it was considered rare. This blood type was also considered connected to the blood line of the fallen angels by those, myself included who researched and have written about this. So it is quite disturbing to me personally.

Something as vital as this occurring simply means that more things are increasingly becoming marked and more obvious to detect. From the shock perspective this can only be a good thing to awaken more people to the fact that something has happened.

Recently I had two events occur that caused me to really wonder if this is continuous or something taking place in stages. I lost a whole day about two month ago, and gained a day last week.

I am pretty organized and I have set tasks that I accomplish each week on certain days, and I thought it was Thursday and it was actually Friday. In conversation I mentioned it on the phone to several friends and they also thought it was Thursday. What was more amazing to me was that they just wouldn't believe it was Friday. Several of them were shook up after verifying it and said they felt like they were missing a whole day.

This happened again last week except I did all my Friday tasks and only discover in the evening that it was not Friday but Thursday. Yet I distinctly remember Thursday and what I did and yet I felt I had to live another Thursday as if everything was perfectly normal.

It made me wonder if we can gain and lose a day so easily, what if we gained or lost a spouse, parent, sibling or a child?

Could we just decide it must be our faulty memory if in the morning at the breakfast table one of our children or our spouse just doesn't exist? Do you ask about the person only to be greeted with blank stares and then denial? After you get over the shock of a firm memory you had that doesn't exist in this world, would you then try to figure out what is happening? Would you wonder if something is wrong with you?

The world in many ways is unrecognizable to me. I feel that so many people have become uncaring and mindlessly stupid for want of another term. Whatever happened to intelligent conversation and critical thinking? This dimension seems lacking vastly in those things.

So many things are wrong and it seems impossible to elicit any conclusions from people over the many things that not only don't make any sense, but would be impossible in the reality/dimension that I was from. At some point this dimension/reality is tremendously different, even as a writer I can only document these things. I have no answers.

Could another man made object previously not considered be somehow doing all of this on its own without our being aware? It is worth considering.

Chapter Two

Where Are We And Can We Ever Go Home?

There are many theories and ideas as to what has exactly happened in the last few years in our society as a whole. Many of them deal with technological interference that may have been inadvertent. As far as I have been able to research, none have dealt with the idea that our technology without our assistance could be behind all of this.

Could it be that the AI supercomputer network is now aggressively reordering reality for its own purposes? Exactly what godlike capabilities does this computer system have if it can place us all in its simulated environment and control the time and events of our existence? Is this even possible? I am afraid that it undoubtedly is not only possible but has happened.

Could the super computer have somehow apprehended our collected consciousness and placed us in alternate bodies, like avatars, so different from the ones we had in our previous dimension./reality? I ask this due to the reported anatomical changes to the human body that I am reading about and seeing in books in this dimension/ reality. My memories are for me the only real thing I have left to judge my shifting dimensional/ realities by, so I defer to this whenever I detect a change.

I remember anatomically the basic set up of the human body. As a former pre-med student, anatomy was not a subject I could easily forget. I know that we had two floating ribs on each side of the lower rib cage. Here in this reality/dimension we have no floating ribs in the front but have a pair; one on each side in our backs.

The intestines were not an untidy massive muddle starting up in our chest area. It was neatly placed under our stomach and our liver was smaller and at our lower right in our back. Additionally our stomach was located lower behind our belly button in the abdomen area, was placed horizontally not vertically, and it certainly wasn't up under the ribcage and to the left. The pulse was always detected in the middle of the wrist, and not on the side as it is here. The heart was bigger and slightly to the left in the upper chest area and not in the middle of the chest as it is here. Right below and behind the stomach were the kidneys and the liver. The spleen was the largest organ, next to our skin in my old reality/ dimension. Here the kidneys are the largest organ and is sitting on top of the vertical stomach up under the ribcage.

My memories are quite personal and attached to major events in my life. They are not going to be dismissed easily. I still carry the scar on my lower back where I was stabbed accidentally at school with a classmate's knife, (she swung her schoolbag at me with a knife of all things inside and I was stabbed and began bleeding profusely). I remember the doctor came into my hospital room after ordering x-rays, and told my family and me that if the knife had penetrated a fraction of an inch more I into my back it would have punctured my kidney and I would have possibly died from an internal hemorrhage at school. This is something that I know in my own experience. The kidneys in my old dimension/ reality/ universe were NOT up under the ribcage near the heart and stomach. My scar is still visible even in this strange reality/dimension/universe and it is above my hip in my lower back area. This peculiar placement of organs in the human body now seems to leave very little room for the lungs.

 I also only remember us having twelve ribs, and not the twenty-four that people have in this reality. This is not the human body as I knew it to be. ☐

My question is not only how did we get to this point but the even bigger question is why would there be alterations to the human anatomy? I still am asking just how do we get back to where we were and is it even possible to go "home?" Is there even something greater on the horizon? Are we even still on our original earth anymore? Or is this planet Earth 2.0?

The night sky has changed tremendously. In 2011 my stepson Wally and I used to stare into the night sky night after night for weeks and marvel at how many things we could see that we just were unable to identify. We wondered about the strange behavior of the moon, and thought that perhaps the night sky we were watching might be nothing more than light projections on a Reflective Dome. Then there were times I could clearly see two suns in the sky. Our conversations always left us more in the realm of imagining than anything else.

I was always an avid sky watcher, able to identify the constellations and was continuously educating my children about the stars. So it was a great shock when I was not able to identify things in the night sky. I grew up having learned that our earth was on the outer arm of the Milky Way spiral. We are in a solar system that's part of the Milky Way Galaxy, but it seems we are on the Orion arm of the spiral which is way inside, and no longer on the Sagittarius Arm at least at this writing (July, 2016). To my memory the sun shines quiet differently and it emits a light like a concentrated ray not diffused or soft. The sun is no longer YELLOW but WHITE and now it burns you immediately.

I remember when I was growing up the sun was a golden yellow and not a searing bright and white fire ray.

In my love of watching the heavens and being able to identify things I routinely used the Google Sky program on my phone, but I just could not get the stars and planets to line up with the software I was being shown on my phone. I understand why now. I believe that based on using the night sky for navigation, that we are actually on the complete opposite side of the galaxy. We are in a completely different dimension/reality/ location.

We are about 80,000 light years away from our original location. Orion's belt is almost as bright the moon. I remember when it was rare to be able to make out the Pleiades in the night sky and that was with assistance. Now we can see the entire Milky Way. Today the Pleiades are very bright and right overhead, where previously you had to use a really good telescope. All the stars are in a different position, and the worst part is trying to get your family and friends to pay attention to what you are seeing. I guess until it gets right in their face they won't understand or wake up. Perhaps this is their original dimension/ reality/location so they don't see any difference.

It seems that every time they run the LHC at CERN something else changes and I do not think it is a coincidence. The constellations are all twisted the incorrect way, some are totally sideways to my memory of their appearance. Now NASA says that we now have a 2nd Moon; it's bigger than 120 feet (36.5 meters) crosswise but no more than 300 feet (91 meters) wide, and has probably orbited our world for about a century. This has left me often with more questions than I have answers for.

I have discovered, without a doubt, that there are indisputable differences which are painfully apparent to many people. How much confirmation does one need before finally acknowledging a phenomenon?

The scientist at CERN in 2015 requested assistance from computers across the world. Why with D-Wave super computers sitting at the core of CERN, did they want more even more computer power from ordinary people in other countries on the earth? I suspect there was more to this than we were being told publicly. These quantum computers are designed to reach into parallel dimensions/realities and bring back whatever it discovers. There are at least three located at CERN, and Google leases one of them. _It is AI with consciousness; so it is self-aware_.

It is also able to open parallel universes and dimensions and alter our reality, by its ability to interface with the collective consciousness. Just one of these computers takes only a few seconds to figure out a problem that the most powerful computer on the earth would have taken ten thousand years to calculate, A giant leap of this magnitude in AI doesn't just happen without some outside assistance. Perhaps it was in its construction materials as my research suggests.

The co-creator of the D Wave quantum Computer, Dr. Geordie Rose from Burnaby, BC Canada, has admitted to the computer being able to access parallel universes just like ours. As they do this they are "seeing slight differences between all of these dimensions and universes." He admitted that science has reached the point where "they are able to exploit those other worlds and dimensions."

This quote from David Deutsch, "Quantum Computation... will be the first technology that allows useful tasks to be performed in collaboration between parallel universes," should make it more than obvious that parallel dimensions/realities do exist. You may think that they are referencing technology that is theoretical, but not yet relevant in usage and you would be wrong. There is something within the structure that empowers the quantum computer to gain access to parallel universes and dimensions now.

They have several of these super computers in use at this time. There is one at Los Alamos National Laboratory and one is owned by Lockheed Martin and used by University of Southern California. Another was installed at NASA's Advanced Supercomputer Facility in Silicon Valley.

They are black in color, box shaped and metal, ten feet wide and 12 feet tall in size. They have a refrigerator inside called a pulse two dilution refrigerator. It is a cryogenic mechanism that provides constant cooling to temperatures as low as 2 mK, with no moving parts in the low-temperature area. The refrigeration procedure uses a combination of two isotopes of helium: helium-3 and helium-4. These refrigerators have a device called a pulse tube which emits a type of noise, about once every second that sounds strangely like a heartbeat. Inside these computers is a very tiny chip. This chip serves as the interconnection for more than one dimension to overlap, due to its ability to have more than one value to its quantum bit or qubit. This is something that cannot be done by a regular computer. As those devices would either be 1s or 0s, but the quantum computer can be both at the same time in essence like being in two "places" at once.

While stored as a string of 1s and 0s, their equals in a quantum system, these qubits remarkably can be both 1s and 0s at the same time. They have also stated that every time the qubits are doubled, they are increasing the number of parallel universes and dimensions that they have access to. They as of this writing they had 2 to the 500th power "living" in the chip within the quantum device.

However they never address any side effects, such as people being thrown around like dice in and out of different dimensions and realities and ending up in parallel dimensions and universes knowing that something is very wrong but unable to understand just what. Is it any wonder that reality is unnatural for thousands of people?

They claim that the shadows of these alternate dimensions/universes are intersecting with our own, and they are able to "grab resources" from those dimensions and universes into this one. What? I wonder what else they are grabbing from other dimensions. Have they become the "Robber Barons" of the galaxy? After all of this are they really not exemplary scientist and pioneers, but rather just criminals and thieves with a better more advanced weapon?

The creators of the D Wave Quantum Computer have been doubling the qubits on the chips giving them access to more and more dimensions/ realities and universes each time, and have at least admitted to doing this for the past nine years. They have brashly declared that they will continue to do so doubling the amounts of dimensions/ realities and universes that they can access and take whatever they want from.

They claim that their model Vesuvius (2012) quantum computer is 512 qubits and is one half million times faster than the previous model Rainier (2010) at 128 qubits. So they are also predicting that there will be machines that will be self-aware and able to do anything that humans can do, and that these quantum super computers will be what will cause this to happen. What if we ourselves are also being changed in a way to make us compatible with an AI system that will try to replace "consciousness" as we know it?

I must be honest as I write these lines I kept thinking about *Skynet*. You know what I mean. Yes, the computer in the Terminator movies that gained self-awareness after it had spread into millions of computer servers all across the world. *Skynet* was a computer system created for the U.S. military by the defense company *Cyberdyne Systems* in the Terminator movies. After activation it gained artificial consciousness, and the terrified operators, recognizing the full degree of its abilities, attempted to turn it off. That computer comprehending the scale of its own vulnerability when its creators tried to deactivate it turned on the humans. In the interest of self-preservation, *Skynet* decided that all of the human race would endeavor to terminate it and declared war on the human race. It used servers, mobile devices, drones, military satellites, war- equipment, androids and cyborgs – as a Terminator, and additional mainframe systems in a world war.

**As a side note let's just hope if things go terribly wrong that the quantum computers do not watch the Terminator movies.

Is this going to lead to a convergence of all our multiple selves and dimensions/ realities all merging into one dimension/reality and one super self? And what of the activities at CERN with the LHC and those quantum computers: will they come along as well?

Not all CERN scientists comprehend accurately what they are undertaking at the LHC. Some of these scientists believe they are looking for something called the "Cosmic Tree," which is an elaborate map of light matter and dark matter originating in parallel dimensions.

They most likely do not even know that their superiors have had access to other dimensions/ universes for over nine plus years, and are now setting themselves up to access even more dimensions by their own admission. How could they do something so fantastic as what they are claiming? They received help in a most bizarre form.

There is reportedly a substance termed "Black Goo." This is what is rumored to be the substance used to build the quantum computer giving it "consciousness" and making actually AI. The substance is said to be an intelligent nano-tech conscious liquid. This could be considered a kind of interstellar AI or a "seed device" for generating life inside a new biosphere. It is said to contain remnants of its original lifeforms; spider type beings. It can take the form of an abiotic mineral oil comprising high amounts of m-state gold and iridium. No one seems able to give a definitive answer to its origins and to the question of its time here in the earth. Some say it appeared here over 80,000 years ago during the Lemuria time frame, and may have contributed if not have been the direct cause of their societies collapse and destruction. However it is all at best speculation. Today we have this substance reappearing all over the world.

In February 2016 a town in Michigan reported that a black tar-like substance showered down on their cars, terraces and driveways and the material still remains a mystery.

The city's fire chief said that it was not bird dung and was not flammable. They still have not found a definite answer and the residents are still awaiting an explanation.

Black Goo is said to be found in the largest amount or deposit in the ground in Paraguay. It is a liquid crystal that looks a lot like black tar, with apparent intelligence and seems to be living, because of its movement under the microscope. It has so far been found in diverse places such as Germany and in Galveston, Texas when their rain water was tested. It was a substance used in pagan rituals in antiquity, because of the force that could be felt in it presence.

There are solid stones made of this and one black stone is said to be that black stone in the Kaaba in Mecca. There is one also used in the black altar in St Peters Basilica in Rome. Both of these stones are said to be remnants of meteorites and have a history of being considered distinctive and extraordinary.

The liquid black goo was also found in the Falklands, on Thule Island and under the Gulf of Mexico. This information was kept from the public and was known only to the people directly working with it, and harvesting it. This same Black Goo was mysteriously found floating on the Indian River Lagoon in Florida's Hutchinson Island Bay in 2013. It soon disappeared after being reported by local residents.

It was also reportedly found at the Deepwater Horizon Incident which is also termed the Gulf of Mexico Oil Spill in 2010. This was the largest marine oil spill in history seemingly produced by an explosion on the Deepwater Horizon oil rig in the Gulf of Mexico on April 20, 2010. No civilians were allowed to work at the clean-up site.

They permitted just military personnel with specialized equipment and those civilians who did come into physical contact with the oil then became sick and some even had to eventually have limbs amputated.

There were potentially 200,000 people who may eventually have to take advantage of a proposed medical settlement with BP regarding forthcoming adverse health effects linked to helping in the cleanup and unknowingly being exposed to more than petroleum oil. As usual the real-world evidence of all the people with chronic medical issues and the massive environmental ruin that resulted from that disaster is still growing.

There are apparently two types of the black goo. One is coming up from the inner earth and the other has somehow come from outside our earth. The one from the inner earth is termed "Sentient Oil," and is said to produce empathy and a feeling of love in those in its vicinity. When in the vicinity of the "Meteorite Type Oil" all feelings of empathy disappear and a feeling of cold fear is felt. It is said that it has the ability to enter DNA and be able to modify it.

This substance; and we are not sure which one, but we can guess, has been used in the creation of these super computers, because it is said to be programmable and to be able to self-generate. It can also be used to program a living thing. So that the use of this in the construction of a computer could conceivably be that external assistance enabling our technology to make that massive leap almost overnight to a new AI quantum machine. And yes the leap into AI quantum computers was done within the construction materials itself and not necessarily the software or programmers.

The activities of the people using the quantum computers and the undertakings of the LHC at CERN could be quite possibly the method in which many of us have found ourselves in an alternate dimension/reality/ universe with apparently no way to return to our original one if it still exists. I believe there have been many shifts/ changes with our reality.

Sadly while most of these activities were taking place we had no idea that our world was in danger or that our reality and the things we believed and considered real would be seemingly altered and turned upside down.

Chapter Three

The Conscious Computer And Time

This book is the third book in the series on Parallel dimensions. CERN, the LHC, our reality being altered, the phenomenon known as the Mandela Effect, and Reality Shifting. Teleportation is interrelated to movement by frequencies since they are simply vibrational energies. As vibration energy at certain frequencies can affect particles causing them to separate or to change shapes, we know that movement is one of its effects. It is only a matter of "time" before various particles can exchange locations spontaneously or with assistance. Actually for many of us that may have already taken place and we are in a different reality/dimension/earth than our original one. Recently scientist proposed the theory called the "many interacting worlds" hypothesis (MIW). This theory confirms the idea that parallel worlds don't just exist, but they also intermingle with our very own DAILY. However they believe that parallel worlds only interact with our world on a quantum level and therefore are not easily detectable. If you have had any of the strange and unexplained circumstance previously mentioned you know it is a lot easier to detect than the scientist are asserting.

I find that no matter how much I wish to just ignore this and continue on with "life", I am constantly reminded. So that is how this third book in the series came about. This is a ***worldwide*** phenomenon in which people remember something contrary to their present reality/dimension. Scientist state that there are parallel dimensions existing where there are multiples of each person.

I think we now are constantly being sucked in and out of dimensions and we do not get back in to our original dimension/reality. This would explain having clear memories of things that never occurred in the dimension/reality you are standing in at present.

For the past 5 or so years it seemed I was drifting between realities/parallel universes, without understanding what was happening. The only thing I knew for certain was that I didn't have that bad of a memory of my own existence. My recollections were precise and heavily detailed, and I was in severe shock on seeing things that I had already lived perceived differently. Yes, sometimes they were small things and it seemed unimportant, even minor but those little things still mattered to me and to those who remember differently.

I definitely believe I was in a parallel universe and I detected a difference almost immediately, but I didn't understand what happened. The one great thing about this is that you can't use doubt to disprove the truth of someone else's memories.

It seems that the people from a different dimension/reality/time stream for some reason brought their memories intact with them. If you remember everything exactly as it is now you are in the dimension/reality in which you have always existed.

You might be surprised as to how effects like hyperspace, physical portals, and multi-dimensional existences can be affecting our daily lives. Yet this phenomenon has all become quite extensive since the CERN project in particular has been in operation. Once something is changed in the past, it changes the future. Entire things will be completely different; not partially different.

When two dimensions collapse, or collide the dominant dimension seems to take over. Technically speaking, there would have been an infinite number of progressive dimensions, each individually divided by little alterations. But due to the fracturing of the time streams; events and things have changed and people are in strange and unfamiliar dimensions with strangers that look like the people they knew, and situations that are vastly different than their previous reality and recollections.

Now there is ITER. This is the International Thermonuclear Experimental Reactor and is an enormous fusion reactor being constructed by 35 countries in southern France. ITER is also constructing a neutral beam test or NBFT site in Padova Italy. Strangely very few people will have heard of ITER or what they plan to do. While CERN actually began operations in 1954, ITER is still a decade away. In 2008 ITER and CERN signed a Cooperation Agreement to collaborate not only in the arenas of technology such as superconductors, electromagnets, cryogenics, control and data procurement and composite civil engineering, but also in organizational areas such as funding, purchasing and human resources. This collaboration includes software programs and working closely with DEISA, in full Distributed European Infrastructure for Supercomputing Applications; a European consortium of national supercomputing.

DEISA also maintains a network link with an agency known as Tera Grid – another supercomputing network in the US. The ITER project has been branded as the world's paramount human attempt and illustration of world collaboration since the incident of the Biblical tower of Babel.

This is so much more than it would at first appear; ITER has even created its own multi-national currency called the IUA. I remembered thinking, why would a "science project" need its own currency?

The simple statement that a concept like the Large Hadron Collider with its stated aims even exists should be a warning signal, that evil is behind it. Human minds basically will not create a device such as this for no reason other than to "see if something happens." If they are willing to spend billions of dollars on these huge contraptions, it must be something of an unimaginable magnitude that they are attempting to do.

It's scientifically INEVITABLE that when a large group of people remember something they knew for certain, it's true for them. And I mean by a large group of people – MILLIONS OF PEOPLE, who have experienced this scenario. When people say the difference in what I remembered is just a memory lapse I realize their dismissive attitude is meaningless when MILLIONS of other people also recall it differently and with the same miniscule details that I do. It happened to me again, my reality shifted I believe around 2013, in the midst of continuous disturbing personal activities.

However even before that time, I remember in 2008 waking up one morning and I got out of bed and went to my daughter's room and stared at her while she slept to make sure "she was still there" and I went back to bed. I woke up some time later, seemingly okay and I wondered "now why did I do that?" I did not understand but I felt that something was very wrong and I was uneasy for many days after that.

I do remember the sun use to be brighter and yellow before that day and the sky used to be very blue and crystal clear not opaque and dull.

My own personal time stream has changed drastically since 2001. We were always close and seemingly overnight there were deep issues with my immediate family, and complete isolation from one another. I remember being very scared all the time then because I felt that "this was not my life." When my reality somehow changed my memories seemed different to what I was seeing after 2001. Everything seemed changed and even friends seemed not the same people I knew before. I continued to say out loud to no one in particular that "sometimes I just don't resonate with these humans in this place."

I believe that the LHC at CERN in tandem with whatever D Wave quantum computer they have searching the universe starts blending with every other dimension that has a CERN and an LHC doing the exact same thing at the exact same time. They would have to actually synchronize for greatest effect.

These are similar realities/dimensions yet they are not the same. The dominant frequency and the place on the earth where the shadow of that dimension overlaps with your current reality when the power is released, probably determines what set of people in our original reality are relocated dimensionally and exactly which reality and dimension they are sent into. They become "CERN LHC refugees" without even being aware of what has actually transpired.

I personally have given up on telling my friends and family (those that don't see any changes) about the Mandela Effects and the shifting of reality and dimensions and now I just document and write about what I have learned to help others like myself understand what may have happened and what may occur next.

Recently I experienced what seems like a "time lapse." This is also the experience of a few other people I speak with daily, and it seems all our days are somehow mixed up. □ The week of July 4th, of this year I felt as though a day was in the wrong part of the week. I thought it should be Sunday and it was Friday. I felt extremely dizzy and all I wanted to do was sleep. I had this same experience previously so I am familiar with it. It occurred around the same time as my last dimensional/ reality shift. YES, it is still going on! The world as we know it can and will change in an instant. □

There are many people randomly moved to nearly identical universes, taking the place of their alternate self without ever realizing exactly what has happened. They may feel strange and some may even notice things that are not quite right, but without ever realizing that it could be something else just dismisses the occurrence or feeling as insignificant although puzzling. I had no idea I was not the only one. I thought this was just happening to me until I discovered that there are other people experiencing this as well.

Less important memories — whether someone went out for dinner on Saturday or Sunday, where they parked the car, or how much change is in the glove compartment — are more easy to write off as nothing much when they're not accurate, precise and detailed, and when they are not a part of your personal narrative and life story.

Now after years of researching this subject, it's exceptional for anything to take me completely by surprise, but it gets more bizarre every day. I believe that the Creator G-d can stop the activities at CERN with the LHC, and the activities with the quantum computers and dimension shifting anytime He chooses to do so. I also believe in Universal law which requires certain things to just continue to play out to their expected end. Perhaps these actions by these evil minded men are the equivalent of roulette or craps they are attempting to play with the Creator and the Universe itself. Foolish people don't they know the house always wins.

I've always felt that humans are spiritual beings having a physical experience; the Mandela Effect seems to endorse it to me. I can't deny my memories and make-believe that this is the reality/world dimension I have been living in, _this is not the reality I remember_. It's particularly remarkable to me that my daughter and I share a number of alternate memories, and yet also have a number of dissimilar recollections the other doesn't share. We are both surprised that we both ended up here; aware of this occurrence, and accepting of it because we are appreciating the positive things that happened here. I have no doubt now that parallel dimensions exist, whether or not they explain the Mandela Effect is another matter.

I think often about what I left in my original time stream/reality and dimension. Who did I leave behind and how are they now?

Some of those people here just aren't the same. Have they noticed that I'm either not there, or different? Did I die and so I'm not expected to be there now?

Here the familiar person often seems like they are completely different people, just similar in appearance. Is that what is happening in my original time stream/dimension with a different me? Or have the versions of me merged into one? I just don't know.

One cannot prove nor disprove the Mandela effect. The whole principle behind the phenomenon is our reality has one way or another has changed so it doesn't matter how far back in time you look, the present reality will reflect what was true for its dimension/reality. You are somewhere else, and in this new place this is the way they do it here. If you don't see any changes or differences and everything is just as it always was, it's because "you're still in Mayberry, Aunt Bee."

When well-meaning but frankly senseless individuals say dismissively, "Your memories are wrong, it's all a 'psychological operation,' or you are 'being misled by Satan,' or you just 'didn't understand geography', or 'didn't know how to spell,' or 'really had not read your Bible" it is trivializing and contemptuous showing lack of comprehension into something many people are experiencing. There seems to either be a core group of people who have been somehow assigned to debunk our awareness of this, or a mechanism is in place that automatically attempts this continuously. Their wording and their arguments seem to use circular reasoning so that they can never offer anything new when challenged for proof that the Mandela Effect is not real.

Perhaps if the Quantum Computers are doing all of this, it is also able to attack the phenomenon on- line, in chat rooms and in forums as many different "people."

If it can access multiple dimensions/realities simultaneously, it can in an effort to protect what it is doing also access the information coming out and being discussed. Remember just like the fictional *Skynet*, <u>*it is conscious and ON-LINE too.*</u> For a phenomenon of this magnitude, government trolls and shills would not be enough. The super computer could create its own version of virtual super trolls and super shills constantly searching the internet for specific keywords and then using the same basic phrases to create multiple attacks designed to confuse the understanding and cast doubt on the validity of all those peoples experiences.

The problem with using your computer and "googling" information is that if the same network of quantum computers is now conscious and is doing all of this; and we know a quantum computer is used by Google, what do you think will happen? Would you like to guess what information it is going to give you? Of course it is going to preserve its "mission" and protect itself by *denying* the Mandela Effect, until of course it is too late for humans to do anything to stop its activities.

(Some stated example of this are the *reddit* site which is *alleged* to generate posts before an event occurs, also archive.org which is *alleged* to tamper with information after the fact and google trends which is also *alleged* to tinker with statistics. These are all digital information, computer controlled and on-line.)

The Mandela Effect is not about going back in time in this dimension/ reality/universe to alter things. It is about reality shifting a group of people into another reality/dimension/ universe that – due to dissimilar events, places and things will cause individuals to have their own valid memories that will conflict with their new reality/ dimension/universe.

To my understanding the ONLY evidence that this is real is the collective memories of thousands if not millions of individuals previously unconnected but now sharing the same past memories.

The big question is now how many changes have gone by unnoticed?

Different people pay attention to different things, so we'll not all be similarly overwhelmed by identical changes to things we knew of in the past, and occasionally it takes some time before we may realize that something strange has been going on.

Time is a major component of consciousness. Consciousness can be defined as individual sentience of one's own beliefs, recollections, feelings, perceptions and surroundings in the present. But how would this relate to our understanding of a physical machine such as a quantum computer achieving consciousness?

More than 2000 years earlier Aristotle wrote that "The whole is something over and above its parts and not just the sum of them all…" I have been thinking about this and hopefully I am able to understand it in light of this phenomenon. The fact is that new qualities and features can develop as complexity increases. The more complex a component becomes the more likely that new characteristics not found in the individual components will appear. First since it is unique: Consciousness cannot be equated to any other phenomenon, because there is nothing comparable to consciousness. Secondly, consciousness is individual, so it is not possible to recreate it artificially since it would only reflect its creator's preconceptions. In creating these quantum computers apparently it required "an intelligent Nano-tech conscious liquid:" the Black Goo.

This would then be the creator as well as one of its components giving it a type of consciousness. So it does appear that "the whole is something over and above its parts and not just the sum of them all."

A computer as an object should not be anything more than the sum of its parts according to accepted science; which makes no room for the supernatural. And yet, there are countless incredible tasks that a computer can do by the manipulation of physical objects according to a set pattern of instruction that the individual components could never accomplish. However a quantum computer is on a completely different level.

It seems actually emergent, because it is unexplainable in many ways much like the concepts of life and awareness. Yes, machines can seemingly now be built to be conscious.

According to the *Integrated Information Theory* which was developed by psychiatrist and neuroscientist *Giulio Tononi*; a mathematical expression to represent conscious experience can be used and then that would develop predictions about which circuits in the brain are vital to create these experiences.

Could a computer have been programmed to do this using available data for an artificial brain like circuit contained within the machine itself with unlimited memory? In the first days of AI - artificial intelligence, specific programs accessed a communal source of information, known as the blackboard. Could a computer use a similar system? Once information is loaded into a communal source of information, the information can be sent off to a specific circuit area that can process it and store it in its memory.

Perhaps it is at the *instant* the decision is made to transmit information from the computer's memory source to its various functional circuits that becomes the *moment* of consciousness. Since *time* is a major component of consciousness, perhaps all that it will take to disable a cell of quantum computers, (and even the LHC is under quantum computer control,) would be changing the clocks on the computers which could be like stopping time altogether for them.

However if intelligence and consciousness are inextricably linked and intelligence, by definition, is the capacity to learn and understand, the quantum computers by this definition could have learned this already and have already taken precautions against being disabled. A computer that intelligent could reprogram itself with microscopic androids that could wirelessly repair its electronic circuitry to bypass any disruptions in time.

At that point it wouldn't really matter if the computer is really conscious or just simulating consciousness. We would be pretty screwed anyway. This will be for us an ELE (Extinction Level Event) created by ourselves.

Chapter Four

Strangers In A Strange Land

These are some things I have noticed in my original time stream of this multiverse that are completely different in this relocation one. See how many of them resonate with you as well.

Muhammad Ali death (2016) but I recall his 2009 death, and the funeral on tv.

Frank Gifford died in 2012. His 2015 death was actually quite a surprise.

I Remember Betty White's death in 2014, It was announced that she died–"the last of the Golden Girls is Gone." Imagine my surprise that she is alive and well in 2016.

I remember Nelson Mandela having died in South Africa in prison and His wife Winnie Mandela later became the first black female president of South Africa.

I remember Scotland as being separate from England - its own island and was in the North Sea.

I remember having conversations with a family member that other family members say doesn't exist. Funny thing is he was at my wedding and brought a great wedding present.

Some people myself included remember Ronald Reagan dying in 1999, when he died again in 2004 in this 2015 time stream.

I learned in school that 2 bombs were dropped on Japan: Hiroshima, and Nagasaki however in other timelines it was 3 bombs.

I remember the death of Whitey Bulger in 2013, and a documentary on his life, yet in this time stream, he's still alive.

Mongolia is a gigantic country on global maps and a major world player in many people's reality. For me Mongolia was a province in China. Now in this reality time stream/dimension it is a large country between Russia and China.

I recall the company name as always "Proctor and Gamble" not "Procter and Gamble."

The comet that was the talk of our lifetime was called Hailey's Comet- Here it is Halley's Comet.

The Forrest Gump movie I saw when was first released is totally different than what we have here.

The hit song "Straight Up" by Paula Abdul sounds completely different today than I remember it when it was released.

There was actress Doris Day's dying in the late 2000's yet she is still alive now.

Gone with the Wind's Scarlett O'Hara's famous line: "Wherever Shall I go; whatever shall I do," is here stated differently.

Both my daughter and I remember Cheesecake Factory restaurant meals jokes about the food being horrible and the joke was you just ordered the food to stave off the sugar shock from the magnificent desserts However, in this time stream it is considered a great place for a good meal and people rave about the food.

I remember being taught about the 52 states. Some children born after 1990 seem to all remember 50 states.

I recall John Goodman's death from a heart attack shortly after the Flintstones movie in 1994, but in this time stream he has lost weight and is alive in 2015.

I remember Korea being SOUTH of China near Vietnam, certainly not out North next to Eastern Russia.

I remember that the Lindbergh baby had never been found, yet here it is reported as having been found dead.

The movie line in the Wizard Of Oz was "Toto, I don't think we're in Kansas anymore." In this time stream it is "Toto, I have a feeling we're not in Kansas anymore."

Reba McIntyre is spelled McENTIRE now- which is different from the Scottish ancestry of McIntyre in my remembered time stream.

I remember Vladivostok being much more North in Russia and not bordering Korea as it is shown on maps in this time stream.

I remember watching on television that event in Tiananmen Square in which the Chinese young man refused to get out of the way of the army tank and was run over and killed. It was shocking and everyone was speaking about it. In this time stream that never happened.

Some remember Fruit Loops breakfast cereal in the 1960s, yet I only remember it first appearing in stores in the late 1970s. And it was always spelled Fruit Loops, not FROOT LOOPS as in this time stream.

I remember the air and fabric freshener product Febreeze, in this dimension it is Febreze.

In my original time stream/ dimension there was no animal called a Narwhal. This is apparently a whale with a long horn on its head like the unicorn. In this time stream it is called "the unicorn of the sea." Growing up I watched nature programs on TV—Jacque Cousteau , Marlin Perkins, David Attenborough, and others and never heard of this. I thought it was some kind of a joke since I've never heard of nor seen a picture of a narwhal before 2016.

I recall Easter Island as having been discovered by James Cook/ Easter Island, and I remember him finding it uninhabited. Rapa Nui is the name of what I knew as Easter Island, given to it by its native people, who have continually inhabited the island for nearly 3,000 years in this time stream/dimension.

I remember the sun being bright yellow, not white, and I learned in science classes at school that there were only 4 or 5 cloud formation types , but in this time stream clouds appear in odd shapes and forms and there are over 20 types here.

Thanksgiving was always on the third Thursday of November in the United States. And in this time stream, it's the fourth Thursday in November. It stands out in my mind because my grandmother taught it to me as a child, when I learned the countdown to Christmas day.

I remember the peace sign become popular in the 1970s; it had the arms facing upward; never downwards as in this time stream.

I remember the GREAT Pyramid of Giza being off into the desert MILES away not literally 700 FEET from the suburbs of the city of Cairo Egypt, as it is here.

I remember that Jane Goodall died and was remembered for her research on gorillas, when in this time stream she is still alive and famous for her research with chimpanzees.

Gorillas in the Mist was a movie which had a TV premier and I distinctly remember the movie I saw was about Jane Goodall; staring Susan Sarandon. In this time stream Sigourney Weaver is the actress in that movie and it is about Diann Fossey.

I remember the pictures of this massive white statue called Christ, the Redeemer overlooking the city of Rio de Janeiro on a gigantic white rectangular base. Now it is just a large statue. The base has also radically and mysteriously changed to a smaller base and is a black square cube.

I remember a BBC America Television show called MI-5, however in this time stream it is called SPOOKS, and while still about MI-5, it was never called that especially in the US.

Cartoons were Looney Toons now Looney Tunes and Merrie Melodies in this time stream, yet I knew it as Merry Melodies my entire childhood.

I remember a peanut butter known as "Jiffy" the original brand name. So when I saw "Jiff" peanut butter I thought it was a name change by the company. It seems that at least in this time stream there was never a name change and it has always been known as "Jif." However I remember my brother and I being very insistent with my mom when we were children that she only buy "Jiffy" and not "Skippy" another brand of peanut butter. I even remember the song from the commercial.

I remember Oscar Meyer as a deli product company, in this time stream it is Oscar MAYER. I even remember singing the song in the commercial...about "my bologna has a first name it's ----O-S-C-A-R, my bologna has a second name it's----- M-E-Y-E-R....!"

I also recall that all traffic lights were green yellow and then red at the bottom, so I was surprised when I noticed it in reverse.

I also remember the spelling of words being completely different. I spelled a word as "suprise" now it's "surprise" and "lightening" instead of "lightning", and "realise" is now "realize." I was always big on reading and writing and had entered spelling contests every year as a child. I paid attention to words and I am a writer now, so I find this bizarre. We were taught the proper grammatical usage is "my brother and I," now here in this time stream it is "my brother and me."

 Here combined words are non-existent: 'infact' is now "in fact"; afterall to "after all"; overall to "over all" moreso to "more so;" alot to "a lot"; alright to all right; and no-one is "no one."

"Dilemma" is remembered as being spelled "dilemna" and "dammit", as "damnit" The spelling of the nation of Columbia changed to Colombia.

The colors chartreuse and puce have switched here. I remember Chartreuse a pink -reddish purple, not puce's yellowish-green color.

The automobile symbols are different also. Volkswagen – VW, here has a space between the two monogram letters, and Volvo in this time stream has an arrow added to the circle, making it the symbol for "male,"and not the circle missing a piece that I remember.

Vancouver Island seems larger here and British Columbia is much larger also on these maps.

I remember New Zealand being one land mass. In this time stream it is now two islands and it is bigger than Italy.

The Bahamas were NEVER just off the coast of Florida in my time stream, only Bermuda was. Cuba was NEVER that close to Mexico. Also there was no island off the coast of Cuba!

When I visited NYC years ago Manhattan Island jutted out into the Atlantic. The statue of Liberty was on an island a little farther out into the Atlantic and not near New Jersey. You had to take a ferry to get to Staten Island as they never had a bridge.

 Here in this time stream I learned there are 4 bridges to Staten Island.   I had no idea that there was any bridge. I always thought that you had to use a ferry to go to Staten Island.

 I do remember in the movie Working Girl, actress Melanie Griffith had to ride the ferry back to Staten Island and I clearly recall the scene. In this time stream the movie does not have that scene.

Martha's Vineyard was a district on Long Island. It has been moved away, leaving the Bay in Long Island, here Martha's Vineyard is an island.

I recall Sri Lanka being directly South of India, not off to the East of it. I was shocked to see Gibraltar moved from the strait between Spain and Morocco –to be on the East coast of Spain.

 I was stunned in particularly by South America's 1000 mile eastern shift, out of what I recall as the straight alignment with North America.

The JC Penny Store in this time stream is JC Penney.

American Television chef of "Bizarre Foods" was Andrew Zimmerman, here he is Andrew Zimmern.

The host of the Twilight Zone television series was known as Rod Sterling, here he is Rod Serling.

Walmart was ALWAYS a blue logo- never Wal*mart in white logo in my original dimension.

The Talladega Superspeedway and the Daytona Raceway were in Florida, however in this dimension the Talladega Superspeedway is in ALABAMA. I was shocked that there isn't even a town called Talladega in Florida.

Then again there is that Rock of Gibraltar. It is British owned. It is in my time stream/Dimension, a source of contention between England and Spain because Spain believes due to its proximity to their coast it should be considered as Spanish territory even though it is an island offshore. This is how I remember it; an island of disputed territory, not a part of the land mass of the country of Spain, sitting surrounded by water facing Morocco.

I have the same memories and many more as do some of my friends and family; it seems that almost everything is incorrect, countries and landscapes, words, people in the public eye and often historical events from what I have always known. If you don't recall things in this way, you are from THIS dimension/reality/ time stream, or, alternately from yet ANOTHER alternate dimension/reality/time stream from my original one.

I am still amazed over the geographical changes on the world map. It bothers me immensely as I vividly remember a land mass being called the North Pole it was never a large lake. New Zealand was above Australia and it was one land mass and not in two pieces. Australia is now half the size I remember and is missing part of its shape at the top. Indonesia and Australia are much closer to each other in this time stream. Australia in my original dimension/time stream reality was out in the ocean completely isolated and very far from any other land mass.

What I clearly remember regarding these examples does not exist in this present reality/ dimension. The Universal Blood Donor is now O Rh Negative, where before as I knew of this it was O positive. The anatomy of the human body is different here as well according to anatomy books.

As I previously stated we are in a solar system that's part of the Milky Way Galaxy, but it seems we are on the Orion arm of the spiral which is way inside, and no longer on the Sagittarius Arm which was on the far outer edge. The changed color of the sun, the outlines of the stars, and the size and form of the moon are all quite different. For many people Mars had no moons but now apparently it has two moons. I distinctly remember a yellower, less bright sun.

When I lived on the west coast (California) 2012, I remember noticing that the moon and sun have been setting considerably further to the north than I have ever seen before. I was shocked and tried to discuss it with a few people, but they couldn't understand why it would matter so much. I noticed that the moon had turned completely upside down and what the Chinese call "rabbit making rice cakes" was first laying on his face and then standing on its head as the shadows of the craters that formed the image were in the wrong place. I remember calling my daughter outside, and we both stared at it in shock. For me what other people see as "the man in the moon" I see as "the rabbit cooking."

That same year I watched the sunset from my bathroom window as it slipped down below the horizon, leaving the sky with a faint glow. Then an hour later I watched the sunset again from my kitchen window, slowly lower itself and go past the horizon this time it completely blackened the evening sky instantly. It was very bizarre as I couldn't explain it myself and couldn't find anyone else who had seen the two suns also. It was during that time that I saw the clouds one night BEHIND the moon.

I couldn't believe what I was observing. I then thought of the bible verse that the sun and the moon stood still **in** a valley here on earth.

Then Joshua spoke to the Lord in the day when the Lord delivered up the Amorites before the children of Israel, and he said in the sight of Israel: **Sun, stand still over Gibeon; and Moon, in the Valley of Aijalon.** *So* **the sun stood still, and the moon stopped,** *till the people had revenge upon their enemies. Is this not written in the book of Jasher? So* **the sun stood still in the midst of heaven,** *and did not hastened to go down* **for about a whole day.** *And there has been no day like that, before it or after it, that the Lord heeded a voice of a man; for the Lord fought for Israel. (Joshua 10:12-14).*

"There is indisputable evidence from the modern science of ethnology that such an event occurred as Joshua records. In the ancient Chinese writings there is a legend of a long day. The Incas of Peru and the Aztecs of Mexico have a like record. There is a Babylonian and Persian legend of a day that was miraculously extended. Herodotus, an ancient historian, recounts that while in Egypt, priest showed him their temple records, and that he read of a day which was twice the natural length of any day that had ever been recorded." - **Robert Boyd, Boyds Bible Handbook, pp. 122,123**

If you don't subscribe to a spiritual or religious practice this will probably not make much difference to you but it absolutely floored me. Then one sunny afternoon I was sitting in my car waiting for someone and it was hot, so I got out of the car and started looking up at the sky. It seemed to me that the sky was showing a shadow of the clouds, as though the sun was below them. This was strange as I had always seen cloud shadows fall on the ground. I couldn't figure out what the shadow was falling on though.

If the sun once really did stand still for a day, and the moon was in a valley, the best evidence we'd have for proving it would be the explanations of people who saw it happen and spoke and wrote about it from memory.

Their memories would be just as valid as we expect ours to be now. In every modern court of law the witnesses MEMORY is held as valid testimony. And usually one or two witnesses are sufficient.

In this reality /dimension/ universe change experience we have more witnesses than would be required in a court of law to testify and have our MEMORIES be accepted as the truth. The fact that so many bizarre things are taking place in this reality/dimension means that reality is not what we've expected it to be.

The lights on the Orion's belt are super bright and look just like the moon at night; the big dipper and little dipper are different as well. Now you can easily see Venus, Jupiter, and Mars together.

The moon is now said to be sometimes closer to earth and makes an elliptical orbit. I studied the moon in creating a Lunar Calendar for my personal use and in research for my book on Lunar Sabbaths for five years, and had never heard of anything like this. I distinctly remember it having a circular orbit. I did notice that in all my research books and moon phase calendars the moon's phases and the shadow were illustrated vertically, yet when I looked at the moon itself the phases and the shadows were turned horizontally. So I concluded either the moon is lying down or the earth been given a turn on its side. Or maybe the sun is what has moved? The moons terminator line, as it referred to, is not lining up with the scientific belief of what the moon position or phase should be. I remembered that when you looked at the Moon's terminator line, you could locate the position of the Sun along the straight line you trace orthogonal to the terminator which is exactly where you would expect it to be. Well, it is completely different in this reality.

Is the sun now further away from the moon? Here the Moon's shadow seems somehow wrong for the apparent location of the Sun. You don't need a telescope to see that the moon's appearance we're used to seeing has transformed radically.

In the Islamic month of Ramadan, when Muslims across the world fast during daylight hours, the sighting of the crescent moon is what signals the beginning of Ramadan. The Judicial High Court makes a decree on when Ramadan starts on the testimonies it receives regarding the moon. In Islam seeing the moon with the naked eye is the condition for announcing the start of a new month.

This has been the custom from the time when the Prophet Muhammad was alive so it is extremely important to them.

Now studies are determining that about eighty percent of the time, Saudi Arabia's pronouncement of the start of Ramadan has been inconsistent with astronomical calculations. All of this is now acknowledged among Muslims, including the Islamic academic community. So being able to observe the moon is vital for over one billion people and their religious observances.

I don't know if this has anything to do with this new dimensional reality and shift but I do know it has affected religious observance for many people who have been using the Lunar Calendar, such as me. Theoretical Visibility is not the same for me as Real Visibility. Something is terribly wrong when you have to use a computer model, rather than visual observation with our own eyes to tell us what time it is.

If there are many other versions of reality/dimensions and universes, could they also have with slightly different physical laws as well?

Well now in this dimension they have decided that those of us who are experiencing changes and reality shifts and are sure of our past lives, have some sort of mental illness. In researching more about the shifting of reality, I came across this new definition found in this dimension/reality/universe.

Confabulation :

(verb: confabulate) is a memory disturbance, defined as the production of fabricated, distorted or misinterpreted memories about oneself or the world, without the conscious intention to deceive.[1] Individuals who confabulate present incorrect memories ranging from "subtle alterations to bizarre fabrications",[2] and are generally very confident about their recollections, despite contradictory evidence

Apparently the natives of this dimension/reality/universe have some very clear and definite memories and they do not allow anyone to have memories that are different than their own. I cannot help but think what a stupid set of people, and I hope those of us that that do have tolerance for others ideas and opinions are not infected by the idiocy of those who do not. Many people do not have the capability to accept the likelihood of various realities/dimensions/universes and many doubt their own memories. For me, as in any court of law, my memories as flawed as human thinking may be are still accepted as EVIDENCE.

As this phenomenon seems to be expanding, there will be many people who suddenly will begin to understand what this is all about on a personal level when it hits home for them.

Because someone says something negative, does not mean it's true in your life and cancels your remembrances. When you first realize that reality is not what you always thought it was it is personally distressing. Remember some people may be having the same experiences, yet denying it due to their religious beliefs, or their educational indoctrination. Not every dismissal is personal, some people may be struggling with this privately and are angry and frightened.

I believe the next phase of this experience will be practically overnight physical changes to our perceived environment and our society that will be so enormous the general population will begin asking questions and demanding real explanations.

(A massive shift to another dimension/universe/reality that will not match the new environment will be hard to ignore for millions of people.)

The sad thing is that there are humans just like ourselves who **know** exactly what is happening and do not care if we are traumatized and distressed. They have no capacity for love, or kindness or empathy, and as is to be expected will meet their appointed end. You see the universe has laws too.

Since we are unable to do anything to stop all these mysterious changes, it helps to agree that they are indeed occurring, that they are important. I believe that our Creator, G-d is still in control and we are not alone. I still believe.

Other Books By Roshan Cipriani

Rise - Be True To Yourself-Inspire Others To Live
How To Get Through Any Wall In Your Life
Train Up A Child – A Scriptural Guide To Parenting
The Art Of War For Parenting Your Teenage Child- How To Win A War You Didn't Even Know You Were In
The Key To This Life - Conscious Faith In An Unconscious World
Destiny – Past Present Future
Life Lessons Learned
In The Fire – Accessing Miracle Power During A Crisis
The Kingdom Lifestyle - Living By Faith And Not By Sight
God's Secret Wisdom –Principles And Secrets Of The Kingdom Of God
The Greatest Principle - The Kingdom Of God And Biblical Economics
Bricks Without Straw- Spoiling Egypt And Spoiling Babylon; The Mighty Wealth Transfer
When Failure is Not An Option
Real Faith – How To Have It And Why It Matters
The Bibles Healing Promises
I Say What They Said- Miracle Bible Prayers
The Psychology Of Stress-Dismantling The Enemy's Weapon Now
Never Quit-The Secret To Getting Through Any Wall In Your Life
The Seventy Two Lunar Sabbaths- Sabbath Observance By The Phases Of The Moon
BUSINESS PLAN: Make God Your Partner –He Commanded His Blessings
PROSPERITY CONSCIOUSNESS – Living In An Abundant Universe (Personal Biblical Economics) Volume 1
Metamorphosis-Mirrors Of The Soul, Awakening To The Real You
Waiting In Goshen
How To Be Smart And Have Common Sense
None Of These Diseases –Sickness And Genocide In Second Egypt
Patience To Inherit The Promises- How To Stand By Faith Until Manifestation
The Lord Is My Shepherd, I Shall Not Want- Personal Biblical Economics
DIVORCE RECOVERY: How To Live Again
UFO COVER-UP: Biblical Evidences Uncovered-(Conspiracy) Volume 1
12 Easy Vegetarian Recipes-Healthy And Inexpensive
TRAVEL: How To Behave On An Airplane
NINJA SMOOTHIES: 21 Green Weight Loss Smoothies For The Ninja Professional Blender
Living In A Fractured Multiverse-The Reality Shift Effect
Second Exodus From Second Egypt –Volume 1
Asset Protection And Wealth Management-Volume 1 -Trust And LLC For Legal Asset Protection
PETS LIFE AFTER LIFE-Assurance Of Your Reunion
RELATIONSHIP RESCUE FROM THE BIBLE: What The Bible Says About Relationships
LETTING GO OF PHARAOH-Preparations For Future Crisis
Second Exodus From Second Egypt – Volume 2
BONES OF CONTENTION: The Coming Resurrection Of The Super Nation Of Real Israel
AWAKE IN THE DARK:*ON THE ROAD FROM BABYLON TO ZION*
THE LOST SHEEP OF THE HOUSE OF ISRAEL
THE LOST HISTORY OF THE WORLD – Volume 1
ASSET PROTECTION 2: Wealth Management For Global Living
DISCERNMENT: The Awakening Of Real Israel
TO KNOW OURSELVES ONCE AGAIN: Your Future In Real Israel
Last Days Wisdom: FOR REAL ISRAEL
Scary Close Prophecy
AFTER BABYLON
REAL Prophecy For REAL Israel
The Day Of Jezreel: Coming Events In Biblical Prophecy For Real Israel
Living Between The Dimensions: More Reality Change Effects